Same Old Horse

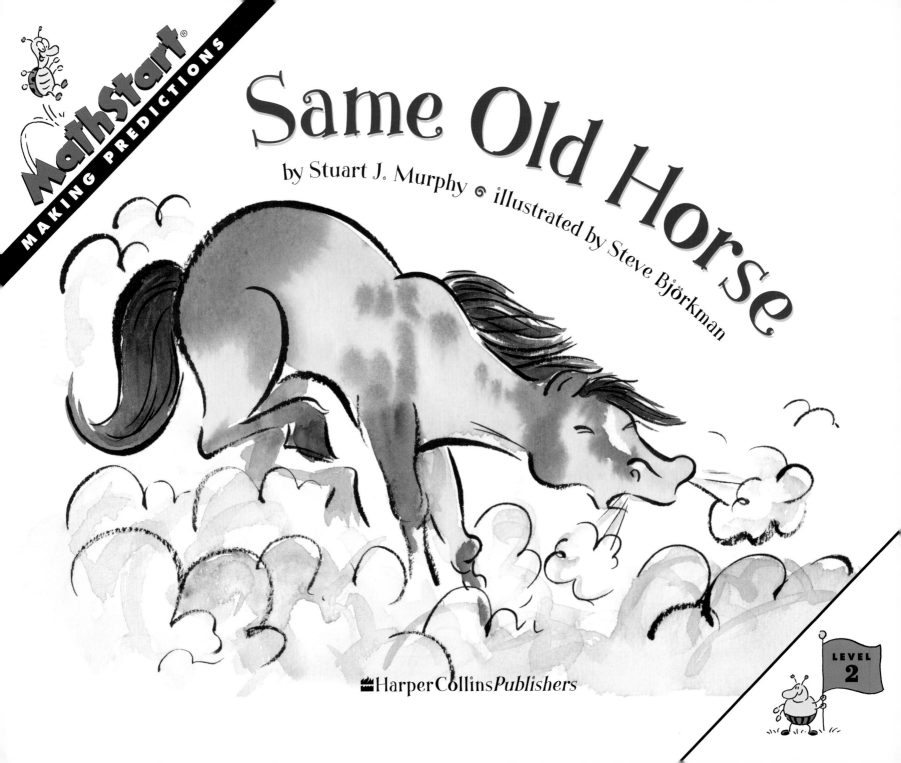

Same Old Horse

by Stuart J. Murphy ● illustrated by Steve Björkman

HarperCollinsPublishers

LEVEL 2

To Jack-be-Nimble and his owner, Michael
—S.J.M.

To Gracie, who loves horses
—S.B.

The publisher and author would like to thank teachers Patricia Chase, Phyllis Goldman, and Patrick Hopfensperger for their help in making the math in MathStart just right for kids.

HarperCollins®, 🎬 ®, and MathStart® are registered trademarks of HarperCollins Publishers.
For more information about the MathStart series, write to
HarperCollins Children's Books, 195 Broadway, New York, NY 10007.
or visit our website at www.mathstartbooks.com.

Bugs incorporated in the MathStart series design were painted by Jon Buller.

Same Old Horse
Text copyright © 2005 by Stuart J. Murphy
Illustrations copyright © 2005 by Steve Björkman
Manufactured in China.

Library of Congress Cataloging-in-Publication Data
Murphy, Stuart J., 1942-
    Same old Horse / by Stuart J. Murphy ; illsutrated by Steve Björkman.—1st ed.
        p.   cm. (MathStart)
    "Making predictions.
    ISBN 0-06-055770-2 — ISBN 0-06-055771-0 (pbk.)
    1. Prediction theory—Juvenile literature.   I. Björkman, Steve, ill.   II. Title.
III. Series.
QA279.2.M87 2005   2004022472
519. 2'87—dc22

Typography by Elynn Cohen   14 15 16 SCP 20 19   ❖   First Edition

Be sure to look for all of these **MathStart** books:

Hankie was allergic to hay. He sneezed about every twenty minutes. That's how he got his name—Handkerchief.

Hankie didn't just sneeze on schedule. Every day he did the same things at the same time.

"Hankie's so predictable," Jazz teased. "We always know exactly what he's going to do."

"Yeah," said Majesty. "What a bore."

Hankie just sneezed—*"AaaCHOO!"*—and walked away.

Hankie hated being teased. But even more, he hated thinking that Jazz and Majesty might be right.

Am I really boring? he wondered.

Hankie's best friend, Spark Plug, tried to make him feel better. "Don't let them bother you," she said. But Hankie didn't listen.

I'll show them I'm not boring, Hankie thought. *I can be as unpredictable as anybody.*

The next day, Jazz and Majesty were hanging around the gate when Spark Plug trotted by.

"Just watch," said Jazz. "Hankie will come out of the barn at exactly ten o'clock."

"You don't know that," said Spark Plug. "You're just guessing."

Jazz winked at Majesty.

Spark Plug didn't know that Jazz and Majesty had been watching Hankie for days. They knew that Hankie's owner, Susan, always brought Hankie out of the barn an hour after she arrived. Today Susan had gotten to the stable at 9:00 A.M.

Inside the barn, Hankie could hear Jazz
and Majesty.

This is my chance to be unpredictable! he thought.
I just have to stay inside until after ten o'clock!

Hankie lowered his head and dug
in his heels. Susan tugged
on the reins.

Hankie sneezed. *"AaaCHOOO!"*

It was such a big sneeze, Hankie couldn't keep his balance.

He staggered through the barn door at exactly 10:00 A.M.

The next day, Spark Plug was chewing on some fresh grass when Jazz and Majesty trotted over.

"I predict that Hankie will be wearing his blue saddle pad today," said Jazz.

WEEK: 1 2 3 4 5 6

Saddle Pad   Red   Blue   Red   Blue   Red   ?

She had noticed that Hankie wore his red pad one week and his blue pad the next. All last week he'd worn the red one.

Inside the barn, Hankie saw a new chance to be unpredictable. He just had to make sure that Susan put on his red saddle pad today.

Hankie snatched his blue pad in his teeth and hid it beneath a big pile of hay.

Then, of course, he sneezed. *"AaaCHOOO!"*

Hay went flying just as Susan arrived.

"What's your pad doing underneath all this hay, Hankie?" she said. And she put the blue pad on Hankie's back.

Oh, no! thought Hankie. *Now I'm as predictable as ever!*

The next day, Jazz and Majesty watched as
Hankie came out into the field to graze.
"I bet he rolls over first thing," said Jazz.

"Of course he will,"
said Majesty. "He's done it
every day for the last five days."

Hankie heard them.

I won't roll over. I won't. I won't! he thought.

But the grass looked so tempting.

Hankie looked over at Jazz and Majesty. They were looking the other way.

Now's my chance! he thought.

He flopped onto his back. He wriggled and rolled. It felt wonderful.

Majesty quickly turned around. "Hah! Same old Hankie!"
he called out. "Same old horse!"

Hankie got to his feet with a sigh.

21

"Now watch," whispered Jazz. "I bet he'll get a drink next. He does that almost every time."

Hankie heard her. But he didn't care anymore. And he was thirsty. He walked over to get a drink.

I am *predictable*, he thought. *I* am *boring*. *There's nothing I can do about it.*

Jazz called out, "I predict that you'll sneeze in three minutes."
She knew that Hankie sneezed about every twenty minutes.
His last sneeze had been seventeen minutes ago.

Spark Plug came over to Hankie.

"Don't worry," she said. "I don't care if you do the same things all the time. You're my best friend, and I like you the way you are."

Hankie felt a little better.

"You're right," said Hankie. "From now on, I'm doing what I want when I want to—no matter what anybody thinks."

And he stuck his nose back into the water.

Just then his sneeze caught up with him.

29

"That was predictable," said Spark Plug.
"But it sure wasn't boring!"

31

In *Same Old Horse*, the math concept is making predictions, an important part of logical thinking. A prediction is not a random guess but rather is based on the observation of a pattern.

If you would like to have more fun with the math concepts presented in *Same Old Horse*, here are a few suggestions:

- Read the story with the child and point out that Majesty and Jazz make predictions about what Hankie will do next. Ask the child to predict what Hankie will do before you read on. Have the child give a reason for his or her prediction.

- Reread the story and have the child predict what Hankie might do the next day or the next week.

- Change the patterns in the charts that are in the book. For example, you might change the chart on page 11 to read:

	Mon.	Tues.	Wed.	Thurs.	Fri.	Sat.
Susan Arrived	9:00	9:15	9:30	9:00	9:15	9:30
Hankie Outside	9:30	9:45	10:00	9:30	9:45	?

Have the child examine each new chart and predict what Hankie will do. Ask the child to explain his or her thinking.

- Have the child ask other members of his or her family to make a list of things they do every day that would make them predictable.

Following are some activities that will help you extend the concepts presented in *Same Old Horse* into a child's everyday life:

School Lunch Menu: Have your child record the school lunch menu for two weeks. Then predict what the menu will be the following week.

Same Old You: For three or four days, have the child keep a chart of certain things he or she does every day. For example:

Activity	Mon.	Tues.	Wed.	Thurs.
Time I got up				
What I had for breakfast				
Color of shirt I wore				
Time I got home from school				

Ask the child if there are patterns. Is the child predictable like Hankie?

The following books include some of the same concepts that are presented in *Same Old Horse*:

- THAT'S GOOD! THAT'S BAD! by Margery Cuyler

- ANNO'S HAT TRICKS by Akihiro Nozaki and Mitsumasa Anno

- THE NAPPING HOUSE by Audrey Wood